Graphing
GRADE 8

Written by
Q Pearce

Illustrated by
Corbin Hillam

Cover Illustration
by Susan Cumbow

FS112042 Graphing Grade 8
All rights reserved—Printed in the U.S.A.

Copyright © 1999 Frank Schaffer Publications, Inc.
23740 Hawthorne Blvd.
Torrance, CA 90505

S0-BCR-438

TABLE OF CONTENTS

Notice! Student pages may be reproduced by the classroom teacher for classroom use only, not for commercial resale. No part of this publication may be reproduced for storage in a retrieval system, or transmitted in any form or by any means—electronic, mechanical, recording, etc.—without the prior written permission of the publisher. Reproduction of these materials for an entire school or school system is strictly prohibited.

INTRODUCTION

This book has been designed to help students succeed in math. It is part of the *Math Minders* series that provides students with opportunities to practice math skills that they will use throughout their lives.

The activities in this book enable students to review the type of information presented in each example, then create similar graphs of their own. The data in each activity is relevant to their everyday lives, or the functioning of their community. This helps students to recognize the purpose behind each example and to perceive the value and practical application of such tools. The activities can be used as supplemental material to reinforce existing curriculum in class or at home.

The skills covered in the book are line plots; single and multiple line graphs; single, double, sliding, and stacked bar graphs; histograms; scatter graphs; circle graphs; Venn diagrams; charts; tables; and time lines. Students will apply other math skills such as addition, subtraction, multiplication, division, fractions, and percents, while interpreting the data.

Graphing

GRADE 8

Name_____

A **line plot** is one of the simplest graphs. It uses a symbol such as an **X** or a dot to represent numerical information. The eighth grade students at Sommerset Jr. High sold magazine subscriptions as a fundraiser. Prizes were awarded to the three top-selling students. This line plot shows how many subscriptions were sold by the top five students.

MAGAZINE SALES

Leann	Raul	Thomas	Maggie	Anne
X				
X				
X	X			
X	X	X		
X	X	X		X
X	X	X		X
X	X	X	X	X
X	X	X	X	X
X	X	X	X	X

Prizes

1st   Boom box

2nd   Radio

3rd   Camera

X = 5 subscriptions

Use the line plot to answer the questions.

A. Who won the boom box? _____

B. How many total subscriptions did the top four students sell? _____

C. What percentage of those sales were Leann's? (Round to the nearest percent.) _____

D. Which two students will not win a prize? _____

E. How many more subscriptions did Leann sell than Raul? _____

F. If the top four students sold $\frac{1}{6}$ of the school total, how many subscriptions were sold altogether? _____

G. Thomas sold $\frac{2}{5}$ of his subscriptions to family members. How may did they buy? _____

H. If the average subscription cost twelve dollars, about how much was Raul's total sales? _____

I. If Maggie sold five more subscriptions, would she have won a prize? _____

© Frank Schaffer Publications, Inc.   Graphing Grade 8

Name_____

iT'S iN THE CARDS ....

Bob's older brother gave him a large collection of sports cards. Create a line plot using the following information to show the different cards in the collection.

There are fifteen baseball cards that are more than twenty years old.

There are five fewer basketball cards than hockey cards.

Half of the baseball cards are less than twenty years old.

There are ten more football cards than basketball cards.

There are twenty hockey cards.

SPORTS CARD COLLECTION

baseball	basketball	football	hockey

X = 5 cards

Use the line plot to answer the questions.

A. How many cards are in the collection? _____

B. Write a fraction to show how many of the total cards are baseball cards. (Write it in lowest terms.) _____

C. How many more football cards are there than hockey cards? _____

D. What percentage of the total cards are football cards? (Round to the nearest percent.) _____

E. 20% of the baseball cards are players on the New York Yankees. How many of the baseball cards are of Yankees? _____

WASH DAY ....

A **bar graph** uses bars to show numerical information. To make sense, a bar graph must include a label for each bar and a scale to show the amount each bar stands for.

Mr. Henry's eighth grade science class wants to raise one hundred dollars for a field trip to the Natural History Museum. They've decided to hold a car wash. This bar graph shows how successful previous car washes have been.

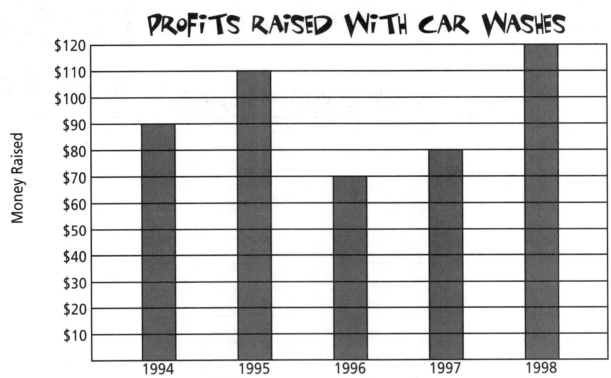

Use the bar graph to answer the questions.

A. Which year were the profits the lowest? _____

B. What was the average amount made per year? _____

C. How much profit was made in 1995? _____

D. Which year showed the greatest increase? _____

E. Which year showed the greatest decrease? _____

F. Is there a predictable trend in sales? _____

G. Are the students certain to reach their goal with a car wash? _____

H. What fraction of car washes made over one hundred dollars? _____

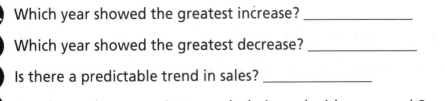

TALL TALES ....

The 10 players on the boy's basketball team are the tallest in their class. Create a bar graph using the information below to show their heights.

Gary is 5 feet 9 inches tall.

Paul and Ray are the shortest by one inch.

Alonzo is one inch shorter than Gary and one inch taller than Nick.

Kevin and Bill are two inches taller than Alonzo.

Jake is 5 feet 8 inches tall.

Craig and Peter are the tallest by one inch.

HEIGHTS OF BASKETBALL PLAYERS

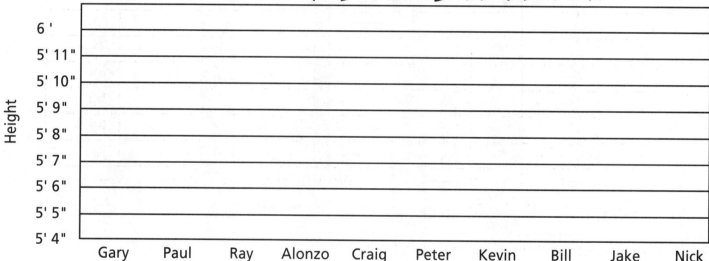

Use the bar graph to answer the questions.

 A. Is Kevin taller than Craig? _____

B. How tall is Raymond? _____

C. How many boys are taller than Gary? _____

 D. What is the average height of the team? (Round to the nearest inch.) _____

 E. What is the difference in height between the shortest and the tallest boys? _____

 F. What percentage of the players are taller than 5 ft 7 in.? _____

 G. Write a fraction to show how many players are shorter than 5 ft 9 in. (Write the fraction in lowest terms.) _____

Name_____

A **double bar graph** enables you to compare sets of information.

The seventh and eighth grade classes at Wellbern Middle School participated in a community beautification project. They planted trees over five weekends. This double bar graph compares their results.

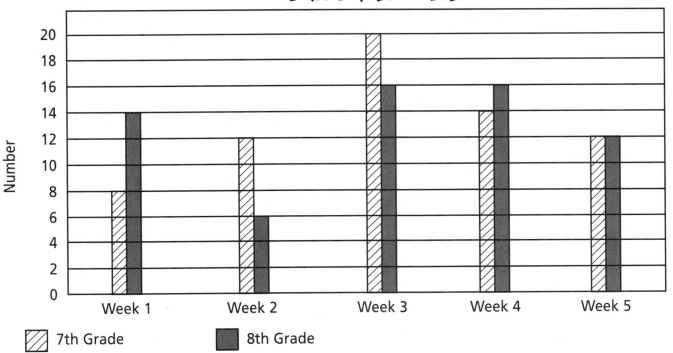

TREES PLANTED

☐ 7th Grade   ■ 8th Grade

Use the double bar graph to answer the questions.

A. How many trees were planted altogether? _____

B. What percentage of the trees did the eighth grade class plant? (Round to the nearest percent.)
_____

C. During which week was the most planting done? _____

D. What was the greatest number of trees planted by one class in one week? _____

E. How many more trees did the seventh grade class plant than the eighth grade class?
_____

F. What was the average number of trees planted per week? _____

G. How many trees were planted during week 2? _____

H. Write a fraction to show how many weeks had less than the average number of trees planted. (Write the fraction in lowest terms.) _____

Name_____

The Jameson Falcons and the Kent Grove Vikings competed in the final basketball game of the season. Create a double bar graph using the information below to show how each team scored per quarter.

Jameson scored 5 more points than Kent Grove in the first and third quarters.

Kent Grove scored 20 points in the second quarter.

Jameson scored 10 points in the fourth quarter.

Kent Grove scored 5 more points in the third quarter than in the second.

Jameson scored 20 points in both the first quarter and the second quarter.

Kent Grove won by 5 points.

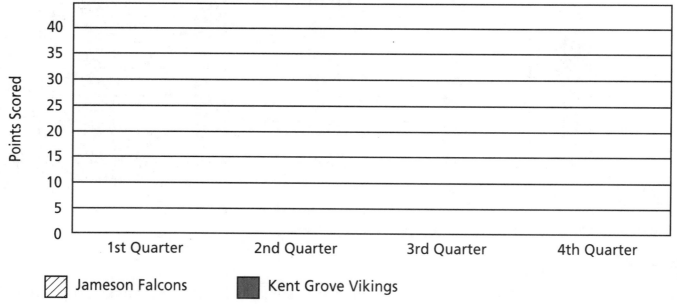

Use the double bar graph to answer the questions.

A. How many points did Kent Grove score in the 4th quarter? _____

B. What was the final score? _____

C. Which was Jameson's highest scoring quarter? _____

D. What was Jameson's average score per quarter? _____

E. What percentage of Jameson's total points were scored in the 2nd quarter? _____

F. In the 1st quarter, what percentage of the total points were scored by Jameson? (Round to the nearest percent.) _____

© Frank Schaffer Publications, Inc.

Name_____

A **multiple bar graph** uses bars of different colors or patterns to allow you to compare sets of information.

The Manchester Beach Middle School offers several student clubs. This multiple bar graph compares three of the student clubs.

CLUB MEMBERSHIP

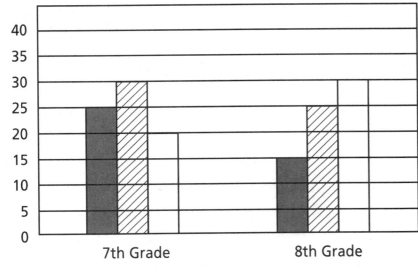

Number

40
35
30
25
20
15
10
5
0

7th Grade   8th Grade

Chess Club   Science Club   Drama Club

Use the multiple bar graph to answer the questions.

A. Which is the most popular club? _____

B. Which club has the most eighth graders? _____

C. What is the total chess club membership? _____

D. Which club do one third of seventh graders belong to? _____

E. In all, how many eighth graders belong to these three clubs? _____

F. What percentage of the drama club is seventh graders? _____

G. Which club has the lowest membership? _____

H. What is the average number of seventh grade students per club? _____

Name_____

GOING THE DISTANCE..........................

Paula, Jason, and Roberto have been training for a bicycle marathon. Create a multiple bar graph using the information below to show their daily mileage.

On day one, Roberto traveled 15 miles.

On day one, Paula traveled 1 mile less than Roberto and 1 mile more than Jason.

On day two, Paula rode 2 miles further than Roberto.

On day two, Roberto rode 8 miles. He rode 1 mile less than Jason.

On day three Paula traveled 2 miles less than she did on day one.

On day three, Jason rode 10 miles. He rode 1 more mile than Roberto.

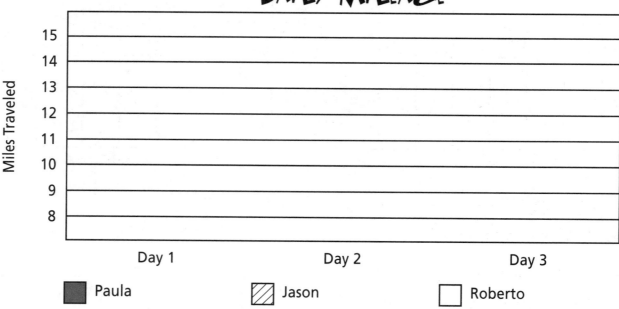

Use the multiple bar graph to answer the questions.

A. Who had the highest total mileage? _____

B. Who had the highest single day mileage? _____

C. What was Paula's average daily mileage? _____

D. What percentage of the total miles Jason traveled were on Day 1? (Round to the nearest percent.) _____

E. Write a fraction to show how many of the total miles Paula traveled were on Day 2. (Write it in lowest terms.) _____

Name_____

A **sliding bar graph** is an alternative to using a double bar graph. It is similar, but the bars are opposite each other rather than side by side.

This sliding bar graph shows the weekly attendance percentages for the seventh and eighth grade students at Dillard Middle School during the month of May.

STUDENT ATTENDANCE IN MAY

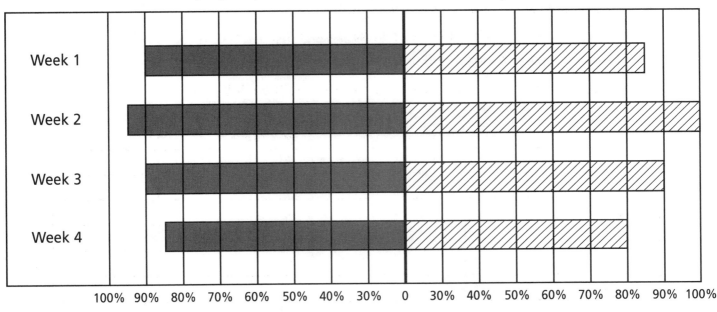

| 7th Grade | | 8th Grade |

Use the sliding bar graph to answer the questions.

A. During which week did the eighth grade students have perfect attendance? _____

B. Which week had the lowest total attendance? _____

C. What was the average weekly attendance for seventh graders? _____

D. Who had the best overall attendance record? _____

E. During which week were 15% of seventh graders absent? _____

F. What is the average weekly absence rate of eighth graders? (Round to the nearest percent.) _____

G. If there are 200 seventh graders, how many were absent during week 1? _____

H. How many eighth graders were absent during week 2? _____

Name_____

A MATTER OF TASTE

Jeffrey decided to do a survey for his science fair project. He wanted to know if girls liked different flavors of juice than boys. He asked 10 girls and 10 boys to taste various flavors of juice and pick the ones they liked.

Create a sliding bar graph using Jeffrey's results.

60% of the boys and 70% of the girls liked cherry.

8 boys and 10 girls liked grape.

30% percent of the girls liked lemon lime and double that percentage of boys liked it.

Half of the boys liked the orange flavor, but only $\frac{1}{10}$ of the girls liked it.

JEFFREY'S SCIENCE FAIR EXPERIMENT

Cherry																				
Grape																				
Lemon lime																				
Orange																				

10 9 8 7 6 5 4 3 2 1 0 1 2 3 4 5 6 7 8 9 10

■ Boys   ▨ Girls

Use the sliding bar graph to answer the questions.

 A. How many boys liked lemon lime? _____

 B. Which flavor did girls like best? _____

 C. Did boys or girls have more evenly distributed preferences? _____

 D. Which flavor was least favored? _____

E. What percentage of girls liked the orange flavor? _____

© Frank Schaffer Publications, Inc.

Graphing Grade 8

Name_____

A **stacked bar graph** can show the same information as a double or multiple bar graph, but by stacking the various bars into one bar, it makes totals easier to read. However, it is harder to tell the total of each individual bar, so they should be numbered.

Eleanor Daly School is having a science fair. This stacked bar graph shows how many entries each grade submitted for each of the different scientific disciplines.

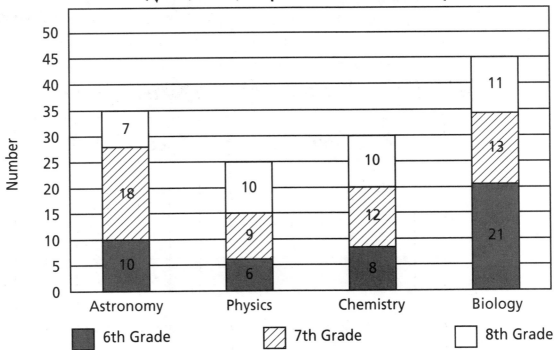

MAKE UP OF THE SCIENCE FAIR

Use the stacked bar graph to answer the questions.

A. How many students participated in the science fair? _____

B. Which scientific discipline is most popular? _____

C. Which discipline is most popular with seventh graders? _____

D. Which grade did the most physics projects? _____

E. Which grade did the fewest chemistry projects? _____

F. If there are 60 students in sixth grade, what percentage participated in the fair? _____

G. What was the average number of chemistry projects per class? _____

H. Write a fraction to show how many of the participating students did a biology project. (Write it in lowest terms.) _____

Name_____

IN DEFENSE OF NATURE ....

The eighth grade students at Davis Jr. High did a survey on the status of endangered species in Davis County over five years.

Create a stacked bar graph using the following information:

In 1994 there were a total of 30 endangered species in the county—15 insects, 4 birds, 4 mammals, and 7 fish.

In 1995, 1 bird, 2 mammals, and 2 fish were removed from the list.

In 1996, 3 insects and 2 fish were removed from the list. The Davis County Wildlife Preserve was created.

In 1997, half of the remaining insects, 2 birds, and 2 mammals were removed from the list.

By 1998, 3 insects, 1 bird, and 1 fish remained on the list.

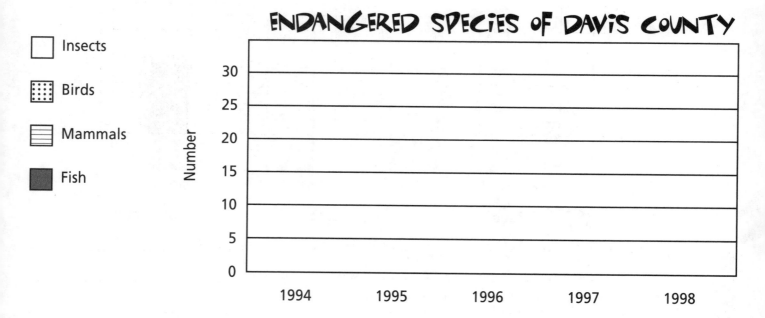

Use the stacked bar graph to answer the questions.

A. Which class of animal was most numerous on the list in 1994? _____

B. Which class enjoyed a 50% improvement in 1995? _____

C. How many animals were removed from the list in 1997 after the Wild Animal Preserve was created? _____

D. Write a fraction to show how many animals recovered and were removed from the list from 1994 to 1998. (Write it in lowest terms.) _____

© Frank Schaffer Publications, Inc.

14

Name_____

EVENING HOURS ....

A **histogram** is a graph that shows the distribution of values in a given population. It is used to show one set of data and is designed for analyzing trends. The vertical axis always stands for a given number of observations or frequency.

The 100 eighth grade students at Jefferson Middle School answered a survey about the average time spent on homework each night. This histogram shows the results of the survey.

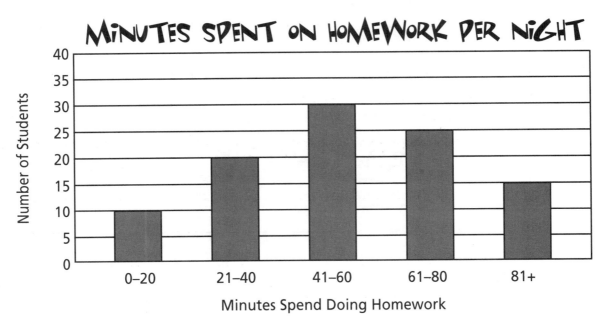

MINUTES SPENT ON HOMEWORK PER NIGHT

Number of Students

Minutes Spend Doing Homework

Use the histogram to answer the questions.

A. How many students spend 20 minutes or less per night doing homework? _____

B. How many students spend more than 80 minutes doing homework? _____

C. What is the most common amount of time spent doing homework? _____

D. What percentage of students spend 21–40 minutes per night doing homework? _____

E. Write a fraction that shows the number of students who spend 41–60 minutes doing homework. (Show it in lowest terms.) _____

F. How many students spend an hour or less doing homework? _____

G. What percentage of students spend more than an hour per night doing homework?

_____

H. How many more students spend 61–80 minutes per night doing homework than students who spend 0–20 minutes doing homework? _____

© Frank Schaffer Publications, Inc.   Graphing Grade 8

Name_____

Mr. Gray's science class conducted an experiment on reaction time. One at a time, each of his students pushed a buzzer when they saw a light flash on. Create a histogram using the information below to reflect the data collected.

Mr. Gray has 50 students.

The reaction times ran from 500 milliseconds (1/2 second) to 1,000 milliseconds (1 second).

Four students responded in 500 milliseconds. Double that number of students responded in 600 milliseconds.

Sixteen students responded in 700 milliseconds.

Fourteen students responded in 800 milliseconds.

Two student responded in 1,000 milliseconds.

The rest of the students responded in 900 milliseconds.

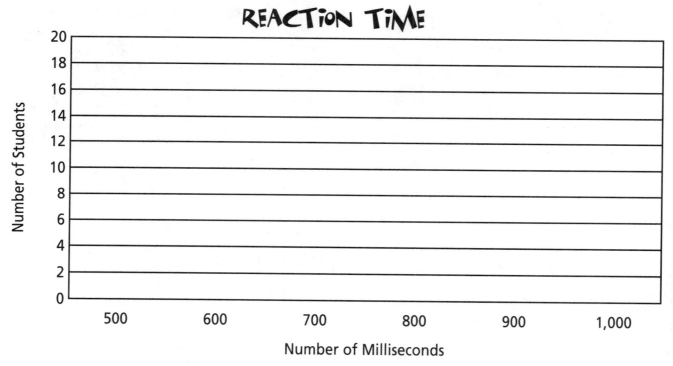

REACTION TIME

Number of Students

Number of Milliseconds

Use the histogram to answer the questions.

 A. What was the most common response time? _____

B. How many students responded at 600 milliseconds or faster? _____

C. How many students responded at 800 milliseconds or slower? _____

D. What was the slowest response time? _____

Name_____

A **line graph** is used to plot information with dots. By connecting the dots, you can see trends in the data collected.

The 30 students in Mrs. Kremer's English class participated in a Read-a-thon. They each agreed to read at least one book per month for five months. This line graph shows the number of books read.

Use the line graph to answer the questions.

A. How many books did the class read in five months? _____

B. How many more books did the class read beyond their original goal? _____

C. What was the average number of books read each month? _____

D. Which months exceeded the average? _____

E. If the average book was 200 pages long, how many pages were read in all? _____

F. If the Read-a-thon continues for a full twelve months, what will the goal be? _____

G. What percentage of books were read in February? (Round to the nearest percent.) _____

H. What percentage of March's goal was met? (Round to the nearest percent.) _____

I. Write a fraction to show how many of the total books were read in January. (Write it in lowest terms.) _____

Name_____

There are 30 students in Mr. Palmer's English class. Create a line graph using the information below to show how many students per month celebrate a birthday during the school year.

Two students have a birthday in September.

An equal number of students have birthdays in October and January.

There are three birthdays in February.

Two students celebrate birthdays in March and three times that number celebrate in January.

Both December and April have three birthdays.

There are no birthdays in November and one birthday in May.

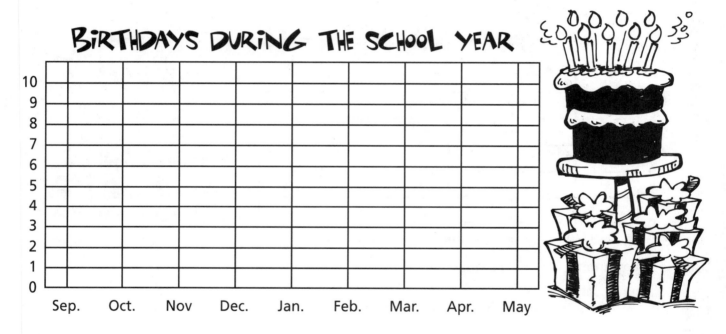

BIRTHDAYS DURING THE SCHOOL YEAR

Use the line graph to answer the questions.

A. How many students had a birthday during the summer? _____

B. How many birthdays were in October? _____

C. In all, how many students had birthdays during the school year? _____

D. During which two months were the most birthdays celebrated? _____

E. What percentage of the students in Mr. Palmer's class celebrate birthdays during the school year? (Round to the nearest percent.) _____

F. What is the average number of birthdays each month during the school year? (Round to the nearest whole number.) _____

Name_____

With a **multiple line graph** you can use different lines to show and compare information.

Marie, Rebecca, and Celine earn money by babysitting on weekends. They each charge five dollars per hour. This multiple line graph compares their earnings.

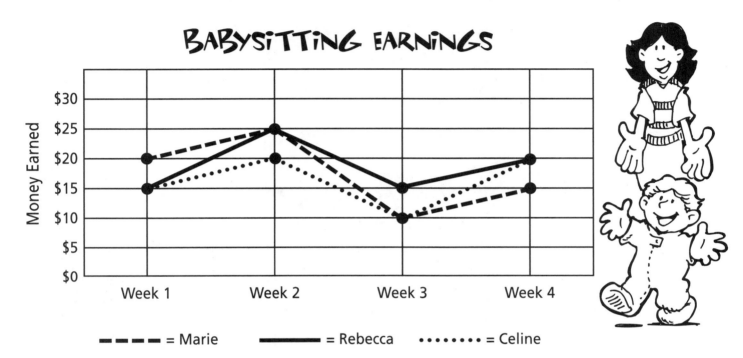

BABYSITTING EARNINGS

= Marie   = Rebecca   ••••••• = Celine

Use the multiple line graph to answer the questions.

A. Who earned the most money? _____

B. How many hours did Celine work during the four weeks? _____

C. What was the average number of hours that Celine worked per week? (Round to the nearest whole number.) _____

D. During which week did the girls work the most hours? _____

E. Who worked the fewest hours in the four weeks? _____

F. How much money did Marie earn in week 3? _____

G. How much money did Marie make altogether? _____

H. What fraction of her total hours did Rebecca work during week 1? (Write it in lowest terms.) _____

I. What percentage of Celine's total earnings were made during week 1? (Round to the nearest percent.) _____

Name_____

There are three main theaters in Garberville. The total price for a movie ticket, a small tub of popcorn, and a small drink has changed over four years. Create a multiple line graph using the information below to show how the prices have changed.

The Star Theater has raised its prices by 50¢ every year.

The price at the Majestic Theater is always 50¢ less than at the Royal Theater.

The Royal Theater charged $6.00 in 1995 and 1996, but raised their price by 50¢ in 1997 and $1.00 in 1998.

The Star Theater charged $4.50 in 1995.

MOVIE PRICES

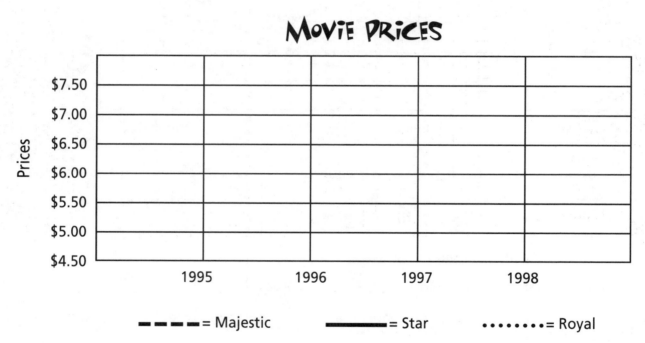

- - - - = Majestic   ———— = Star   • • • • • • • = Royal

Use the multiple line graph to answer the questions.

A. Which theater has the lowest prices in Garberville? _____

B. What was the average price in 1996? _____

C. If the ticket price makes up $\frac{1}{3}$ of the total cost, how much was popcorn and a soda at The Star Theater in 1995? _____

D. Which theater had the smallest price increase between 1995 and 1998? _____

E. If 500 people purchased a ticket, a small popcorn, and a small drink at The Royal Theater each day in 1997, how much did they make per day? _____

Name_____

A **scatter graph** is used to plot two variables against one another.

The 12 students of Mrs. Bridges after school science class are studying ecology. To see how effective her classes are, she has decided to give a test before a week of lessons and after a week of lessons and graph any improvement. This scatter graph shows the results.

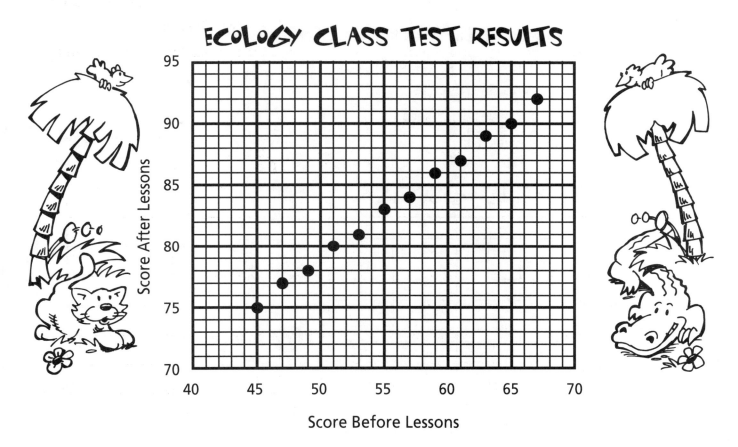

Use the scatter graph to answer the questions.

A. What was the lowest score before the lessons? _____

B. What was the highest score after the lessons? _____

C. How many points did the student with the lowest score before the lessons improve after the lessons? _____

D. After the lessons, what was the score for the student who scored 55 points before the lessons? _____

E. How many points did the student with the highest score before the lessons improve after the lessons? _____

ON THE RUN ·····················

Scott is training to make the track team. His time in the 100 yard dash is 15 seconds, but he needs to cut that to 13 seconds. Create a scatter graph using the following information to show his progress over 10 weeks of summer training.

Before the first week of training, Scott starts out at 15 seconds.

In week 1, he improves his time by .7 seconds.

In week 2, he improves his time by .4 seconds.

In week 3, he improves his time by .3 seconds.

In week 4, he improves his time by .2 seconds.

In weeks 5, 6, and 7, he shaves off .1 seconds.

He achieves his goal in week 10.

TRAINING RESULTS

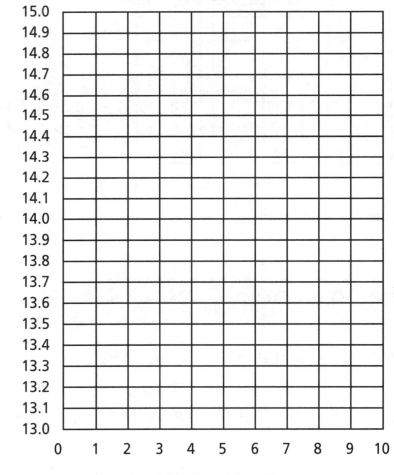

Weeks of Practice

Use the scatter graph to answer the questions.

A. During which week did Scott make the most improvement?

_____

B. In total, how many seconds did he cut off his time in 10 weeks?

_____

C. If this rate of improvement continues for 2 more weeks, will he run the 100 yard dash in less than 12 seconds? _____

D. By which week had Scott's time improved by 12%?

_____

E. Write a question about the graph for a classmate to answer.

_____

_____

_____

_____

Name_____

A **circle graph** shows how sets of numerical information relate to each other.

Diane received a new computer for her birthday, but she will have to purchase her own supplies. This circle graph shows how she will spend her money.

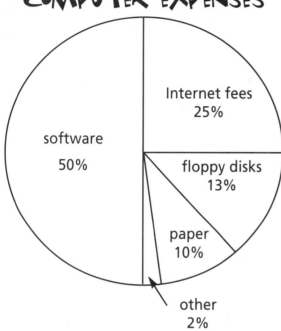

COMPUTER EXPENSES

Internet fees 25%

software 50%

floppy disks 13%

paper 10%

other 2%

Use the circle graph to answer the questions.

 A. What is Diane's greatest expense? _____

 B. On what does Diane spend 10% of her money? _____

 C. If Diane has $200.00, how much will she spend on floppy disks? _____

 D. How much will Diane spend on Internet fees? _____

 E. What percentage has Diane put aside for other expenses? _____

 F. On what does Diane spend five times more than she spends on paper? _____

 G. The amount Diane has to spend on supplies is $\frac{1}{10}$ the cost of the computer. How much was the computer? _____

H. If Diane spends 50% less on software, how much more will she have for other expenses? _____

Name_____

Adam did a survey of his eighth grade math class to find out what types of music the students listen to most often. Create a circle graph using the results below from his survey.

Adam surveyed 30 students.

In all, 15 students listen to rock music.

There are 3 students who listen to jazz.

Twice the amount of jazz listeners prefer rap music.

There are 3 students who listen to country music.

10% of the students enjoy classical music.

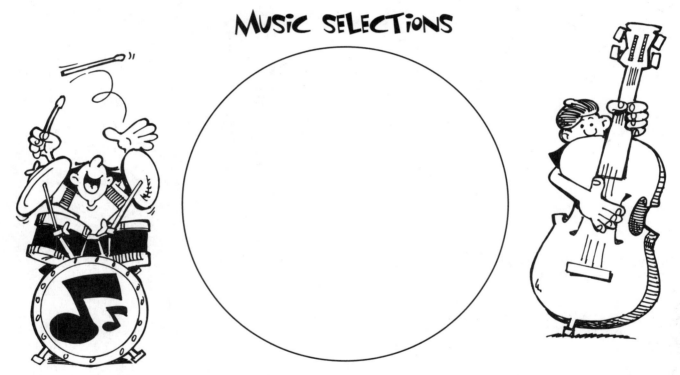

MUSIC SELECTIONS

Use the circle graph to answer the questions.

A. What percentage of students listen to rap music? _____

B. Which is the most popular type of music? _____

C. How many students prefer classical music? _____

D. Which type of music did six students choose? _____

E. Write a fraction to show how many of the students prefer country music. (Write it in lowest terms.) _____

Name_____

A **Venn diagram** uses circles to show relationships between sets of data and how the information may overlap.

This Venn diagram shows the favorite types of transportation of the 60 students at Grover Jr. High.

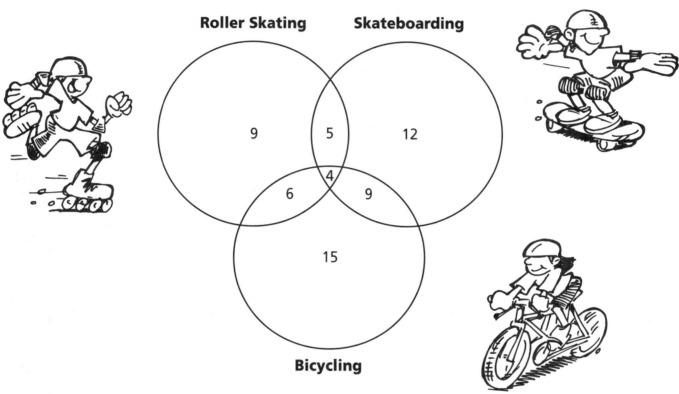

FAVORITE TRANSPORTATION

Use the Venn diagram to answer the questions.

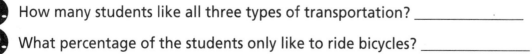

A. Which is the most popular form of transportation? _____

B. In all, how many students like to skateboard? _____

C. How many students like all three types of transportation? _____

D. What percentage of the students only like to ride bicycles? _____

E. Write a fraction to show how many of the students like both roller skating and skateboarding, but not bicycling. (Show it in lowest terms.) _____

F. Which is the least popular type of transportation? _____

G. What percentage of students do not like to skateboard? _____

H. Which type of transportation was selected as the only choice for 15% of the students? _____

Name_____

LEISURE TIME ....

Create a Venn diagram using the information below to show the three main ways the student's in Mr. Carter's math class like to relax.

There are forty students in the class.

Seven students relax only by watching TV.

Five students enjoy both reading and listening to music.

10% of the students only listen to music, while twice that number like to watch TV and listen to music.

Three students only like to read

Six students enjoy both reading and watching TV.

The rest of the students enjoy all three ways of relaxing.

HOW STUDENTS RELAX

Watching TV   **Listening to Music**

Reading

Use the Venn diagram to answer the questions.

 A. How many students only watch TV? _____

 B. How many students participated in this survey? _____

 C. What percentage of students only listen to music? _____

 D. Write a fraction to show how many students enjoy both reading and watching TV, but not listening to music. (Write it in lowest terms.) _____

 E. What percentage of students do not enjoy listening to music? _____

 F. What percentage of students enjoy all three ways of relaxing? (Round to the nearest percent.) _____

Name_____

A **population table** uses rows and columns to organize a specific type of information. It is used to show the population of a given area over time. This population table shows the student population of the four middle schools in Crescent City over five years.

CRESCENT CITY STUDENT POPULATION

	1994	1995	1996	1997	1998
Valley	200	253	262	255	300
Ridgeway	187	203	245	255	286
E.L. Miller	304	317	298	312	304
Vista Verde	268	275	278	289	295

Use the population table to answer the questions.

A. Which school had the same population in 1994 and 1998? _____

B. Which school showed an overall increase of 50% from 1994 to 1998? _____

C. In what year did Valley and Ridgeway have the same number of students? _____

D. What was the increase at Vista Verde between 1994 and 1996? _____

E. Which two schools had decreases in student population during the 5-year period?

_____

F. What was the population of Ridgeway in 1995? _____

G. Which school has the smallest number of students? _____

H. In which year was the population of E.L. Miller below 300? _____

I. What was the average number of students per year at Vista Verde? _____

J. What was the average number of students per year at E.L. Miller? _____

K. Write a question about the table for a classmate to answer. _____

_____

Name_____

Create a population table using the following information to show the population growth of the four largest cities in Hernando County over four years. Remember, since you are showing a population of people, you can only work with whole numbers. Round each answer to the nearest whole number.

Donner had a population of 310 in 1995. It increased by 10% each year through 1998.

Visata City had a population of 85 in 1995. It doubled its population in 1996, went up by 50% in 1997, and doubled again in 1998.

Claymoor's 1995 population of 200 increased by 100 in 1996, by 55 in 1997, and by 45 in 1998.

Wheaton had a population of 210 in 1995, 340 in 1996, 325 in 1997, and 395 in 1998.

POPULATION OF HERNANDO COUNTY

	1995	1996	1997	1998
Donner				
Visata City				
Claymoor				
Wheaton				

Use the population table to answer the questions.

A. Which town doubled its population between 1995 and 1998? _____

B. Which town had a one time drop in population? _____

C. Which town had the fastest rate of growth? _____

D. What was the population of Donner in 1996? _____

E. What is the average yearly population of Claymoor? (Round to the nearest whole number.) _____

F. Which town had a 50% growth in population from 1996 to 1997? _____

G. Write a question about the table for a classmate to answer. _____
_____

Name_____

A **time line** is used to organize events into an easy-to-read order.

The eighth grade students of Frasier Jr. High are having a spring dance. The dance committee has eight weeks to prepare everything. Follow the time line to see their plan.

PREPARATIONS FOR SCHOOL DANCE

Week 1	Reserve multipurpose room
Week 2	Hire disc jockey
Week 3	Choose theme
Week 4	Prepare decorations
Week 5	Mail flyers
Week 6	Order refreshments
Week 7	Put up posters
Week 8	Sell tickets

Use the time line to answer the questions.

A. When do tickets go on sale?

_____

B. What is the first preparation?

_____

C. What happens during Week 3?

_____

D. When are the first flyers sent out?

_____

E. When do the posters go up?

_____

F. What happens in Week 6?

_____

G. How many weeks before the tickets are sold, are the decorations prepared? _____

H. Is the disc jockey hired before or after the flyers are sent out? _____

Name_____

MY LIFE ·

Think about the important events in your life. Then create your own time line. Begin with your birthdate and end at the present.

IT'S MY LIFE

_____ |
(date of birth)

© Frank Schaffer Publications, Inc.   Graphing Grade 8

ANSWERS

Page 3
A. Leann
B. 135
C. 33%
D. Anne and Maggie
E. 10
F. 810
G. 12
H. About $420
I. no

Page 4
Graph should be completed as shown:

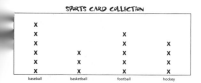

A. 90
B. 1/3
C. 5
D. 28%
E. 6

Page 5
A. 1996 E. 1996
B. $94 F. no
C. $110 G. no
D. 1998 H. 2/5

Page 6
Graph should be completed as shown:

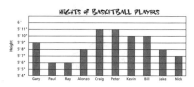

A. no E. 5 inches
B. 5' 6" F. 70%
C. 4 G. 1/2
D. 5' 9"

Page 7
A. 130 E. 2
B. 49% F. 26
C. Week 3 G. 18
D. 20 H. 3/5

Page 8
Graph should be completed as shown:

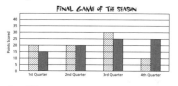

A. 25
B. 85 to 80
C. 3rd
D. 20
E. 25%
F. 57%

Page 9
A. Science Club E. 70
B. Drama Club F. 40%
C. 40 G. Chess Club
D. Chess Club H. 25

Page 10
Graph should be completed as shown:

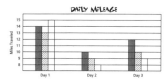

A. Paula
B. Roberto
C. 12
D. 41%
E. 5/18

Page 12
A. Week 2 E. Week 4
B. Week 4 F. 11%
C. 90% G. 20
D. 7th grade H. 0

Page 12
Graph should be completed as shown:

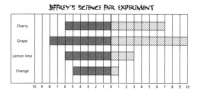

A. 6
B. grape
C. boys
D. orange
E. 10%

Page 13
A. 135 E. 6th
B. Biology F. 75%
C. Astronomy G. 10
D. 8th H. 1/3

Page 14
Graph should be completed as shown:

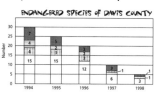

A. Insects
B. Mammals
C. 10
D. 5/6

Page 15
A. 10
B. 15
C. 41–60 minutes
D. 20%
E. 3/10
F. 60
G. 40%
H. 15

Page 16
Graph should be completed as shown:

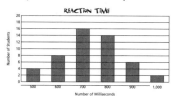

A. 700 milliseconds
B. 12
C. 22
D. 1,000 milliseconds

Page 17
A. 210 F. 360 books
B. 60 G. 19%
C. 42 H. 83%
D. April and May I. 1/7
E. 42,000

© Frank Schaffer Publications, Inc. Graphing Grade 8

ANSWERS

Page 18
Graph should be completed as shown:

A. 4
B. 6
C. 26
D. October and January
E. 87%
F. 3

Page 19
A. Rebecca   F. $10
B. 13   G. $70
C. 3   H. 1/5
D. Week 2   I. 23%
E. Celine

Page 20
Graph should be completed as shown:

A. The Star
B. $5.50
C. $3.00
D. The Star
E. $3,250

Page 21
A. 45   D. 83
B. 92   E. 25
C. 30

Page 22
Graph should be completed as shown:

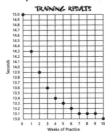

A. Week 1
B. 2 seconds
C. no
D. Week 6
E. Questions will vary.

Page 23
A. software   E. 2%
B. paper   F. software
C. $26   G. $2,000
D. $50   H. $50

Page 24
Graph should be completed as shown:

A. 20%
B. rock
C. 3
D. rap
E. 1/10

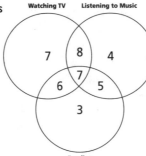

Page 25
A. bicycling   E. 1/12
B. 30   F. roller skating
C. 4   G. 50%
D. 25%   H. roller skating

Page 26
Venn diagram should be completed as shown:

A. 7
B. 40
C. 10%
D. 3/20
E. 40%
F. 18%

HOW STUDENTS RELAX

Watching TV   Listening to Music

7   8   4
7
6   5
3

Reading

Page 27
A. E.L. Miller
B. Valley
C. 1997
D. 10 students
E. Valley and E.L. Miller
F. 203
G. Ridgeway
H. 1996
I. 281
J. 307
K. Questions will vary.

Page 28
Table should be completed as shown:

POPULATION OF HERNANDO COUNTY

	1995	1996	1997	1998
Donner	310	341	375	413
Visata City	85	170	255	510
Claymoor	200	300	355	400
Wheaton	210	340	325	395

A. Claymoor
B. Wheaton
C. Visata City
D. 341
E. 314
F. Visata City
G. Questions will vary.

Page 29
A. Week 8
B. Reserve multipurpose room
C. A theme is chosen.
D. Week 5
E. Week 7
F. Refreshments are ordered.
G. 4 weeks
H. before

Page 30
Time lines will vary.

© Frank Schaffer Publications, Inc.

Graphing Grade 8